Animal Life Cycles

Frog

by Elizabeth Neuenfeldt

Blastoff! Readers 2

Bellwether Media • Minneapolis, MN

Blastoff! Readers are carefully developed by literacy experts to build reading stamina and move students toward fluency by combining standards-based content with developmentally appropriate text.

Level 1 provides the most support through repetition of high-frequency words, light text, predictable sentence patterns, and strong visual support.

Level 2 offers early readers a bit more challenge through varied sentences, increased text load, and text-supportive special features.

Level 3 advances early-fluent readers toward fluency through increased text load, less reliance on photos, advancing concepts, longer sentences, and more complex special features.

★ **Blastoff! Universe**

Reading Level

Blastoff! Beginners — Grade K → Blastoff! Readers — Grades 1–3 → Blastoff! Discovery — Grade 4

This edition first published in 2021 by Bellwether Media, Inc.

Library of Congress Cataloging-in-Publication Data

Names: Neuenfeldt, Elizabeth, author.
Title: Frog / by Elizabeth Neuenfeldt.
Description: Minneapolis, MN : Bellwether Media, 2021. | Series: Blastoff! reader : Animal life cycles | Includes bibliographical references and index. | Audience: Ages 5-8 | Audience: Grades K-1 |
Summary: "Relevant images match informative text in this introduction to the life cycle of a frog. Intended for students in kindergarten through third grade"-- Provided by publisher.
Identifiers: LCCN 2020036794 (print) | LCCN 2020036795 (ebook) | ISBN 9781644874097 (library binding) | ISBN 9781648340864 (ebook)
Subjects: LCSH: Frogs--Life cycles--Juvenile literature.
Classification: LCC QL668.E2 N48 2021 (print) | LCC QL668.E2 (ebook) | DDC 597.8/9--dc23
LC record available at https://lccn.loc.gov/2020036794
LC ebook record available at https://lccn.loc.gov/2020036795

Editor: Betsy Rathburn Designer: Jeffrey Kollock

Printed in the United States of America, North Mankato, MN.

Table of Contents

What Are Frogs?

Frogs are **amphibians**. They can live in water and on land!

Most female frogs lay eggs in lakes, ponds, or **vernal pools**.

Life Cycle of a Frog

Frogs lay their eggs in **clutches**. One clutch can have thousands of eggs!

Some types of frogs take care of their eggs. Other frogs do not.

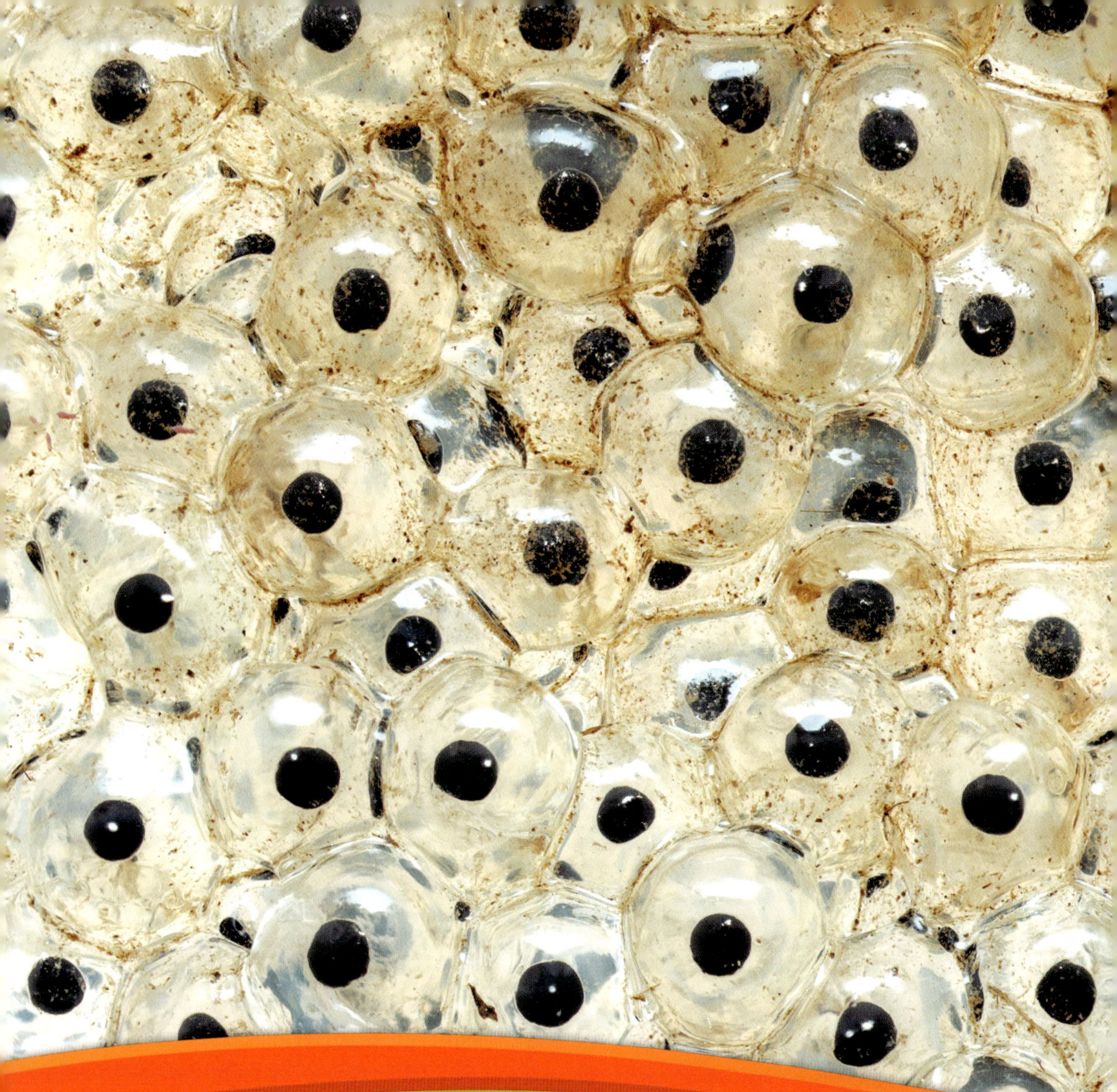

Frog eggs are tiny and soft. It takes one to four weeks for eggs to **develop** and **hatch**.

Growing Up: Wood Frog

- **Egg** → 7 to 28 days
- **Tadpole** → about 2 to 4 months
- **Froglet** → about 1 to 3 years
- **Adult** → about 3 to 5 years

Then **metamorphosis** begins!

When the eggs hatch,
baby frogs are born.
They are called **tadpoles**.

Tadpoles have **gills** and tails. They can only live underwater.

For their first week, tadpoles eat their eggs. Afterwards, they swim to find **algae**.

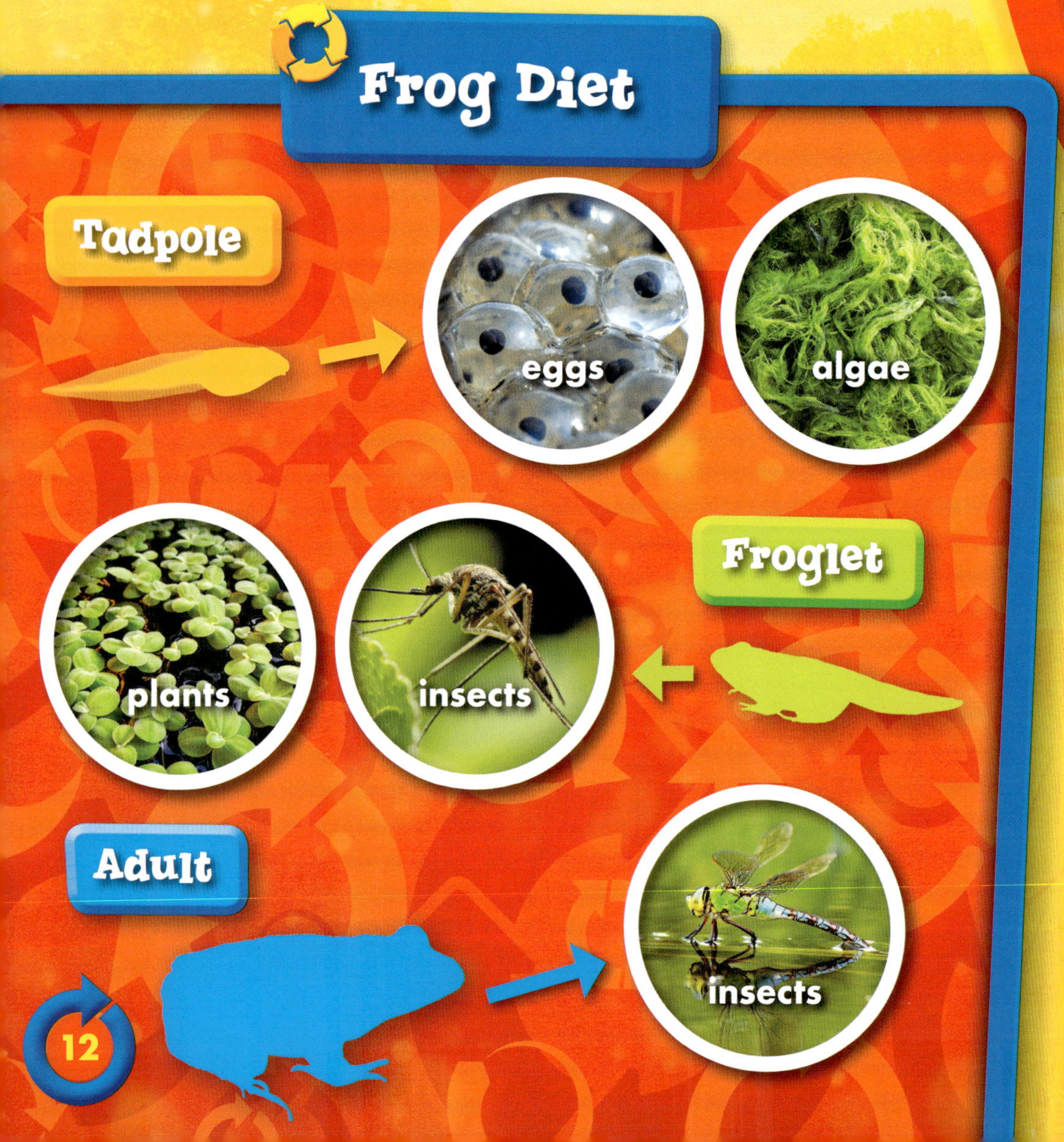

Some tadpoles swim together in **schools**. There may be hundreds of tadpoles in a school!

Most tadpoles grow for two to four months. They develop legs and **lungs**. Their heads change shape. Now they are **froglets**!

Froglets can walk and breathe on land.

froglet

Froglets get food on land or in water. They eat plants and **insects**.

Soon, froglets' tails begin to shrink. This stage can last weeks or even years!

Full-grown Frogs!

Once their tails are gone, froglets become adult frogs.

Adult frogs live on land. They eat insects and live in groups called **armies**.

In the spring, frogs return to the water to **mate**. Male frogs croak in search of females. Soon, the frog life cycle will begin again!

Life Cycle of a Frog

Glossary

algae—plants or plantlike living things that usually grow in water

amphibians—animals that can live on land and in water

armies—groups of adult frogs

clutches—groups of eggs laid at one time

develop—to grow

froglets—young frogs

gills—body parts that allow animals to breathe underwater

hatch—to break out of an egg

insects—small animals with six legs and bodies divided into three parts

lungs—body parts that people and animals use to breathe air

mate—to join together to make young

metamorphosis—a big change in how some animals look and act, usually related to growing

schools—groups of tadpoles that swim together

tadpoles—newborn frogs; tadpoles live in water and have gills, mouths, and long tails.

vernal pools—bodies of water that last for a short period of time and are not very deep

To Learn More

AT THE LIBRARY

Kelly, Irene. *A Frog's Life*. New York, N.Y.: Holiday House, 2018.

Tonkin, Rachel. *Egg to Frog*. New York, N.Y.: Crabtree Publishing Company, 2019.

Vasilyeva, Anastasiya. *Frog*. New York, N.Y.: Bearport Publishing, 2017.

ON THE WEB

FACTSURFER

Factsurfer.com gives you a safe, fun way to find more information.

1. Go to www.factsurfer.com.
2. Enter "frog" into the search box and click .
3. Select your book cover to see a list of related content.

Index

The images in this book are reproduced through the courtesy of: Chros, front cover, p. 1; Eric Isselee, p. 3; Durden Images, pp. 4, 5; IRINA ORLOVA, p. 5; Dan4Earth, p. 5 (vernal pool); Clara Bastian, p. 6; David Tipling Photo Library/ Alamy, pp. 6, 7; Gannet77, pp. 8, 9; Suwat wongkham, pp. 10, 11; Esa Hiltula, p. 11; blickwinkel/ Alamy, p. 11 (gills); Marius Neacsa, p. 12 (eggs); moonoi172, p. 12 (algae); TorD, p. 12 (plants); Achkin, p. 12 (insects top); R0macho, p. 12 (insects bottom); Nigel Cattlin/ Alamy, pp. 12, 13; Macrolife, p. 14; Martijn Strietman, pp. 14, 15; Margaret Welby/ Alamy, pp. 16, 17; jiade, p. 16; SIMON SHIM, p. 17; imageBROKER/ Alamy, pp. 18, 19; Kurit afshen, p. 20; Pixeljoy, p. 21 (egg); Steve Byland, p. 21 (tadpole); Michael Benard, p. 21 (froglet); Gerald A. DeBoer, p. 21 (adult); Adam Gryko, p. 23.